YOUR KNOWLEDGE HAS VALUE

- We will publish your bachelor's and master's thesis, essays and papers

- Your own eBook and book - sold worldwide in all relevant shops

- Earn money with each sale

Upload your text at www.GRIN.com and publish for free

Umar Sambo Umar

Geology and Structural Analysis of Foliation Planes of Granite Gneiss exposed around Kanwara, Dass. Bauchi State

GRIN Publishing

Bibliographic information published by the German National Library:

The German National Library lists this publication in the National Bibliography; detailed bibliographic data are available on the Internet at http://dnb.dnb.de .

Imprint:

Copyright © 2011 GRIN Verlag GmbH
Print and binding: Books on Demand GmbH, Norderstedt Germany
ISBN: 978-3-656-84602-4

GEOLOGY AND STRUCTURAL ANALYSIS OF FOLIATION

PLANES OF GRANITE GNEISS EXPOSED AROUND KANWARA,

DASS. BAUCHI STATE.

BY

UMAR SAMBO UMAR

(05/17644/1)

A THESIS SUBMITTED TO GEOLOGY PROGRAMME, SCHOOL OF

SCIENCE,

ABUBAKAR TAFAWA BALEWA UNIVERISTY, BAUCHI.

IN PARTIAL FULFILLMENT OF THE REQUIREMENT FOR THE

AWARD OF DEGREE, BACHELOR OF TECHNOLOGY (B.TECH)

HONOURS IN APPLIED GEOLOGY

AUGUST 2011

Dedication

Dedicated to the memory of my late grand father Alhaji Umar Jumberi. May his soul continue to rest in Peace, Ameen.

Acknowledgement

All praise is to Allah, the Beneficent, the merciful, and the Cherisher of the universe. I thank him for the thought I am induced with, the charisma, the courage, the wisdom, the enthusiasm that warrant my successes, of which this project is among.

Most sincere appreciation to my parents, Alhaji Sambo Jumberi & Hajiya Dada Sambo, for their prayers, support, cares, and assistance. In fact everything they been doing for me.

My appreciation to my supervisor Mal. Ahmad Isah Haruna for the assistance, advises and for his effort by guiding us to the field.

Thanks to all staff of Geology Programme ATBU, especially the Programme Co-ordinator Dr. A.S maigari and staff; Prof. E.F.C. Dike, Prof. D.M.Orazuluke, Associate Professor N.K.Samaila, Dr. M.B.Abubakar, Malam Amadu Tukur, Malam M.T.Isah and Mr. T.P.Bata.

I also appreciate your cares and concern, my dear brothers and sisters Hajiya Amina Sambo, Hajiya Aisha Sambo, Muhammadu Sambo, Hadiza Sambo, Asma'u Sambo, Sadiq Sambo, Maryam Sambo, Aliyu Sambo and my wife Hasana Ibrahim.

My dear friends and colleagues at home and in school, I sincerely acknowledge your prayers and cares all this while, especially my co-workers Abdul-Smart and Nazif Umar.

Abstract

This work produces some vital geologic information on the geology and structural analysis of rocks mapped around Kanwara, Dass local government area of Bauchi State. The mapped area is located between longitude $N10^001'09"$ and $N1O^000'09"$ and latitude $E9^032'42"$ and $E9^031'26"$. Exposure of rocks in the area were mostly high level and medium level outcrops with only a few low level outcrops, the high level outcrops were conically shaped with height ranging between 400m to 700m above sea level while the inselberg was about 900m. The predominant rock types in the area were granite gneiss, biotite granite and migmatite, under thin section, minerals were identified via their respective optical properties under both plane polarized light and cross polarized light, the major minerals associated with the rocks were; biotite, quartz, feldspars, (orthoclase & plagioclase), muscovite and microcline. Based on field observations, the geologic structures found on the rocks include; joints, foliations, minor folds, ghost schist, dykes and veins. The joints indicate the masking effect of the pan African orogeny on the older granites (trending NW-SE). The shape of the girdle from the structural analysis conducted indicates that the rocks of the mapped area were subjected to medium grade metamorphism.

TABLE OF CONTENT

Preliminary pages   **Page**

number

Cover page……………………………………………………………….i

Declaration……………………………………………………………ii

Dedication……………………………………………………………iii

Acknowledgement……………………………………………………iv

Abstract………………………………………………………………vi

Chapter one

1.0 Introduction……………………………………………………….13

1.1 Aim of study………………………………………………………13

1.2 Location and accessibility…………………………………………14

1.3 Relief and drainage………………………………………………..15

1.4 Climate and vegetation……………………………………………15

1.5 Settlement and land use……………………………………………15

Chapter two

2.0 Review of the geology of the Precambrian to lower

Paleozoic rocks of northern Nigeria................ 17

2.1 The Paleozoic and Precambrian rocks in northern Nigeria............................18

2.2 Intermediate rocks from older granite complex...................................21

2.3 Field relations in Dass..23

Chapter three

3.0 Materials and Methods..25

3.1 Field equipment...25

3.2 Sample collection...26

3.3 Measurements of strikes and dips..26

3.4 Study of rock samples...26

3.5 Procedures for structural analysis (stereographic projections)................28

Chapter four

4.0 Macroscopic and microscopic studies...31

4.1 Macroscopic study of samples A..33

4.1.1 Microscopic study of sample A ([plate I).....................................33

4.1.2 Statistics of sample A (plate I)...35

4.2 Macroscopic study of Samples B..35

4.2.1 Microscopic study of sample B ([plate II)....................................36

4.2.2 Statistics of sample B (plate II)………………………………………38

4.3 Macroscopic study of Samples C……………………………………….39

4.3.1Microscopic study of sample C ([plate III)……………………………40

4.3.2 Statistics of sample C (plate III)………………………………………41

4.4 Macroscopic study of Samples D……………………………………….42

4.4.1Microscopic study of sample D ([plate IV)……………………………42

4.4.2 Statistics of sample D (plate IV)………………………………………44

4.5 Macroscopic study of Samples E……………………………………….45

4.5.1Microscopic study of sample E ([plate V)……………………………46

4.5.2 Statistics of sample E (plate V)………………………………………47

4.6 Summary of the petrography……………………………………………47

4.7 Structural geology………………………………………………………48

4.8 Structural Analysis………………………………………………………53

Chapter five

Results and Summary………………………………………………………67

Conclusion…………………………………………………………………70

Geological Map of the study area…………………………………………72

Recommendations…………………………………………………………73

Reference…………………………………………………………………74

List of Figure   **Page number**

Figure 1……Geological map of Bauchi state…………………………..…14

Figure 2a…..Map of Nigeria showing the Nigerian basement complex…….22

Figure 3…... Granitic and intermediate rocks of Dass area……………..……24

Figure 4……Schmidt net…………………………………………………54

Figure 5….....S-Pole diagram for migmatites………………………….…..56

Figure 6….....S-Pole diagram for gneisses………………………………...58

Figure 7….....S-Pole diagram for biotite granite…………………………...60

Figure 8….....S -Pole diagram for all the 73 attitude readings……………..61

Figure 9……Counting net…………………………………………………...63

Figure 10 ….Points plotted within Hexagons………………………………64

Figure 11…..Contoured diagram……………………………………...….66

Figure12…...Development of S-Pole during folding……………………...69

Figure13…..Geological Map of the study area………………………..……72

List of Tables   **Page number**

Table 1…….Paleozoic-Precambrian rocks of northern Nigeria…………..20

Table 2……Statistics of sample A……………………………..………………35

Table 3……Statistics of sample B……………………………..………………38

Table 4……Statistics of sample C………………………………………………41

Table 5……Statistics of sample D……………………………………………..44

Table 6……Statistics of sample E……………………………………………47

List of Plates   **Page number**

Plate 1……...Photograph of hand sample A…………………...33

Plate 2……. Photomicrograph of sample A…………….......33

Plate 3……...Photograph of hand sample B………………...36

Plate 4……. Photomicrograph of sample B……………........37

Plate 5……...Photograph of hand sample C……………….....39

Plate 6……...Photomicrograph of sample C…………………40

Plate 7……...Photograph of hand sample D…………………42

Plate 8……...Photomicrograph of sample D…………………43

Plate 9……...Photograph of hand sample E…………………45

Plate 10……Photomicrograph of sample E…………………46

Plate11……Tree growing in a joint within granite gneiss…..49

Plate12……Foliations within granite gneiss………………...50

Plate 13…..Minor folds within granite gneiss………………51

Plate14……Pegmatite dyke hosted by biotite granite………52

Plate15……Quartz vein within biotite granite………………52

CHAPTER ONE

<u>INTRODUCTION</u>

1.1 Aims and objectives

1. To produce some vital information on the Geology of the area around kanwara, Dass. This   information comprises the structural Geology, major rock types following evidences from petrography, field relations, and mode of occurrence, colours and observable mineralogy of the rock samples.

2. To suggest the type and level of deformation in the area using stereographic projections.

3. To produce the geologic map of the area, this comprises of the major rock types and topography. Cross section on the map will then be produce which further provides clarification on the geology of the study area.

4. To review some of the works earlier done in the area around 1960 to 1980 and to present, hence builds on what is known.

5. In partial fulfillment for the award of degree Bachelor of Technology (B.TECH) Honours in Applied Geology.

1.2 Location and Accessibility

Though the field work was done during the beginning of the raining season, the area was quite accessible. The road linking Bauchi-Dass passes through NW to SE part of the portion (see figure 1, below), and footpaths and cattle tracks crosses all over, linking nearby villages. However, climbing the outcrops was very tedious as most of the outcrops were conically shaped.

Figure 1, Geological map of Bauchi State showing the location of the mapped area (Adapted from B.S.A.D.P. 1983).

1.3 Relief and Drainage

The area has appreciable relief, as the outcrops were not uniformly shaped ranges are; 600m, 650m, 700m and 720meters above sea level. The inselberg may rich up to 900m, based on deduction from topographic map.

The drainage pattern was therefore, dendritic type of stream which form branches (tree like) and directed by the ridges of the outcrops. The streams drains to a River called (River Sala).

1.4 Climate and Vegetation

The climates of the area consist of a rainy season (extending from May/June to October/November) and a dry season which was characterized by the Harmattan from November/December to April/May.

The vegetation type is Savannah composed of scattered trees, Shrubs and mainly flat lying grasses. The grasses along the river band tend to be greenish all year round (according to one peter who is living there, I did not see it during dry session).

1.5 Settlement and Land Use

The major land use in the area is farming and cattle rearing (Grazing) in some place not used for farming activities. The product of rock weathering in the area provided fertile soil and the drainage systems provided soil moisture. These two factors made the area agriculturally lively and very viable. Along the river bands, Fadama farming like tomatoes were planted and rice while in other places cereals are planted which are mostly maize and guinea-corn.

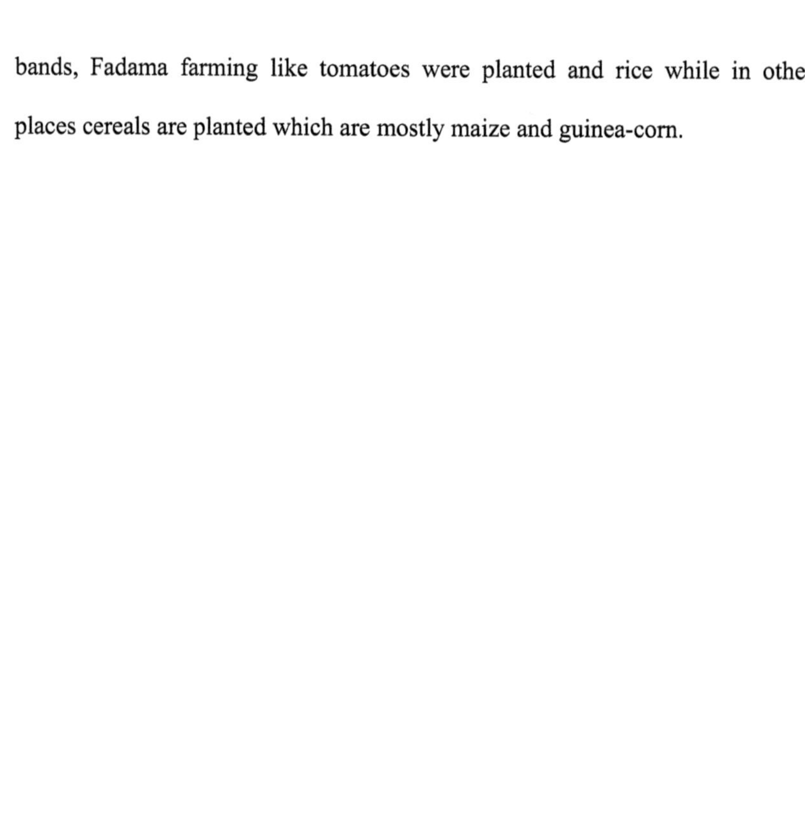

CHAPTER TWO

<u>**LITERATURE REVIEW**</u>

2.0   Review of the Geology of the Precambrian to lower Paleozoic rocks of northern Nigeria

Northern Nigeria is underlain by gneisses, migmatites and metasediments of Precambrian age which have been intruded by a series of granitic rocks of late to lower palaeozioc age. The oldest rocks are represented by a series of older metasediments and gneisses believed to be of Birrimian age and older. These rocks have been variably metamorphosed and granitised through at least two tectonic – metamorphic cycles so that they have been largely converted to migmatites and granite – gneiss. Younger metasediments, believed to be upper Proterozoic in age, were deposited on this granitised basement and filled along with it during the pan – Africa orogeny. They are of low metamorphic grade and are now represented as synclinal trough among older rocks in north – west Nigeria. Intrusive into both the basement and the younger super crustal cover is a series of basic, intermediate and acid plutonic rocks known as the Older Granites. The youngest rocks in the area belong to a suit of volcanic rocks intruded into Older Granite bodies during Lower Palaeozioc epirogenic uplift following the Pan – African orogeny

2.1 The palaeozoic and precambrian rocks in northern Nigeria can be divided into four major groups:

2.1.1 **The Basement Complex (Senso Stricto);** underlies the entire area and includes all rocks older than the Late Palaeozioc metasediments. It includes metasediments of high metamorphic grade such as paragneiss, basic and Calcareous Schist, marble and quartzite, as well as orthogneiss and possible early (Ebunian) granite. The whole Basement complex has been through at least two tecto – metamorphic cycles, and consequent metamorphism, migmatisation and granitisation has extremely modified the Original rocks so that they generally occur as relief rafts and xenoliths in migmatites and granites. More extensive areas of gneiss and ancient metasediments recognized within the Basement complex are believed to be reliefs of an old super crustal cover of probable Birrimian age, and are termed older metasediments in order to distinguish them from late palaeozioc super crustal sediments which are known as the Younger metasediments.

2.1.2. **Younger Metasediments;** From well – defined, approximately north – south trending belts which are extremely developed in the north – west. They are not recorded east of longitude 8^0 but it is believed that they represent the remnants of a once more extensive super crustal cover. The rocks types are varied, ranging from psammitic to pelitic sediments of low metamorphic grade, with occasional conglometric Facies and inter bedded lavas. They have been steeply folded along with the Basement complex during the Pan – Africa Orogeny, so that they now occur in synclinarial keels in a sea of granitic material, resembling the greenstone migmatites association of cratonic regions (Wright and McCurry 1970a).

2.1.3. The Older Granite Series; Involves rocks intruded during the Pan – African orogenic cycle. The earliest rocks appear to be small, irregular bodies of quartz and pyroxene diorite, and gabbro. In the north – west, south basic rocks have been extensively intruded by later acid rocks so that they now form part of an acid – basic complex. Granitisation on an extensive scale profoundly modified earlier rocks and resulted in widespread migmatites and granitic rocks. This granitisation culminated in the intrusion of a diverse and widespread suite of syn to late – tectonic granites, granodiorites and syenites.

2.1.4.   Younger Volcanic Rocks; are the youngest rocks recognized in both north – west and north – east Nigeria. They belong to a post – older Granite episode of high – level magmatic activity. Preliminary age determinations suggest that these rocks were intruded during epirogenic uplift and fracturing in the final stage of the Pan – African Orogeny. For this reasons, they are included in a description of the 'basement' rocks and not with the other volcanic rocks of Nigeria.

Age	Succession	Rock Types
Ordovician – Cambrian	Maradun Group, Kisemi Group, Burashika Group	Granites, porphyries, tuffs, rhyolites, dacites, andesite, basalts.
Cambrian – upper Proterozoic	Older Granite	Granite, granodiorites, syenites, migmatites acid-Basic complex, pyroxene-diorite.
Upper Proterozoic	Younger Metasediments	gabbro phyllites, Schist, quartzite, conglomerates interbedded volcanic
Older Precambrian	Older metasediments and gneiss	Quartzite, marble, basic and calcareous Schist, biotite and hornblende gneiss, granolite

Table 1, Generalized Succession of Paleozoic – Precambrian rocks of Northern Nigeria (After McCurry, 1972.)

2.2 Intermediate rocks from older granite complex of the Bauchi area, northern Nigeria

Previously unmapped area around Bauchi (Survey sheet number 149) have been mapped and some areas previously mapped by Basin (1926) and Oyawoye (1958, 1961, 1962) have been reinvestigated. The area consist of amphibolite Facies Basement migmatites and biotite gneiss with ancient metasediments remnants, occasionally of granulite Facies, and older Granite complexes of acid and intermediate rocks – Bauchite and quartz diorite of Bauchi and Bununu Dass (hearafter referred to as Dass) are described, and some new data and interpretations are presented.

The unusual fayalite quartz monzonite at Bauchi town was first described as a coarse grained augite syenite by Falconer (1911). Basin (1926) also described as syenite from Bauchi, but the distinctive features of the rocks were first emphasized by Oyawoye (1958, 1961) who named it Bauchite (1965). There are four main occurrences of Bauchite in the area, at Bauchi, Yelwa, margas and south of Kangere. In earth case biotite hornblende granite forms a large part of the complex, and quartz diorites are also common. Fresh specimens of bauchite are dark green due to the green or brown colour of quartz and feldspar, the most conspicous crystal being twinned alkali feldspar up to 5cm long. Bauchites usually are small, approximately circular.

Quartz diorites are found in and near to the intrusions (1km across) or as dykes in the bauchite. They are dark green and vary in grain size, but the larger intrusions, e.g. Inkil Hill, have feldspars up to several centimeters in length and appear rather similar to bauchite. All diorites are quartz bearing and most contain at least 10% hypersthene. In bauchite areas diorite forms topography similar to the bauchite, but elsewhere the only endure diorite may be the characteristically spherical, deeply weathered boulders scattered over flat, low – lying ground often adjacent to impressive hill, as at Dass and Kir.

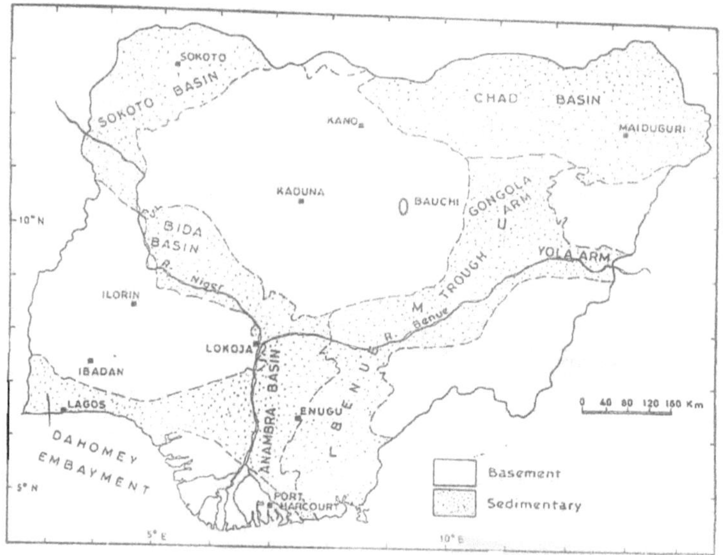

Figure 2a, Map of Nigeria showing the Nigerian Basement comlpex

2.3 Field relations at Dass

At Dass area of diorite forms the eastern edge of an older Granite complex of various types of biotite granite and biotite hornblende granite. Although the hornblende is less abundant than biotite in the latter, unlike the granites associated with bauchite. Rocks of bauchite composition and mineralogy have been found at Dass. The western side of the complex has been described by E.P Wright (1971) and is shown on the Geological survey of Nigeria map sheet 140 (Toro) and 169 (Maijuju) as undivided 'porphyritic biotite and biotite – hornblende – granite'. The main Dass diorite has field relationship with surrounding rocks in some ways similar to those of bauchite with biotite hornblende granite. Diorite is the more basic core of the complex, and grades outwards into a less mafic rock.

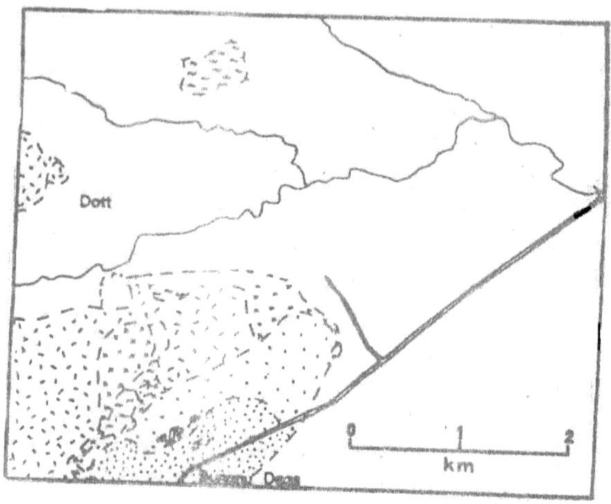

Figure 3, Granitic and intermediate Rocks of Dass Area (After E.P.Wright, 1971)

CHAPTER THREE

MATERIALS AND METHODS

Reconnaissance survey was done in the mapped area, as I was mapping along the flanks of the outcrops and on the rocks, fresh rock samples were obtained and it was completed by studying the samples petrographically in the laboratory and the geological map was also produced to explain the lithology of the rock types encountered. Hence the Methods used in this study are divided into two; Field Methods and Laboratory Methods

3.1 Field Methods

Field methods explain the procedure for sample collection, sample preparation and the equipments used for the field work.

3.1.1 Field Equipments

The field equipments used were

- Geographic positioning system (GPS)

- Geologic Compass

- Geologic Hammer

- Field measuring tape

- Camera

- Field note book

- Sampling Bag

- Topographic Sheet (Sheet 169, Maijuju).

3.1.2 Sample Collection

Fresh samples of different rock types were obtained from the field, because weathered samples can hardly yield us any information.

These samples were labeled (the location number) bagged and brought to school for further analysis (petrography).

3.1.3 Measurements of strikes and dips.

Geologic compass was used in measurements of strike and dip of the foliation planes and of lineaments such as pegmatite dykes.

3.2 laboratory methods

Laboratory methods explain the laboratory procedures for thin section and stereographic projections.

3.2.1 Microscopic Study of Rock Samples

The rock samples were studied macroscopically and microscopically. Macroscopic study involves studying the rock samples without the aid of the hand lens, the colour and the texture were all observed and the samples were named based on field data.

Microscopic study of the rock samples is done in laboratory using the petrographic microscope. In petrography, the minerals were identified via their optical properties under both plane polarized light and cross polarized light.

3.2.1(a) Plane polarized light

Under plane polarized light, the following properties were identified

- Colour of mineral

- Cleavage

- Form

- Relief

3.2.1(b) Cross polarized light

Under cross polarized light the following properties were observed.

- twinning

- interference colours

- extinction

3.2.2 Procedure for generating data for statistical analysis

1. The slides used for petrography are mounted onto the microscope stage

2. The minerals which are already identified are then counted at the first position

3. The stage of the microscope is then rotated which moves to the second position and then the minerals are also counted as they appear under the microscope

4. These procedure is repeated to the third, fourth up to the eight or tenth position

5. The percentage of each mineral is then calculated in relation to other minerals identified. These, is considered to reflects the overall percentage of each of the minerals in the bulk rock.

3.3 Procedures for structural analysis using stereographic projections

3.3.1 The materials needed are;

- Schmidt net

- Counting net

- Over lay tracing sheets

- Pencils

- Cleaners

- Razor blades

- Plane sheets

- Office pins

- Masking tapes

- Rotring pens and ruler

3.3.2 General Procedures for plotting of foliation planes

1. with an overlay sheet in place (laid on the Schmidt net) make a small mark over the point of the Schmidt net

2. To locate the line of strike count off clockwise from North and make a small mark over the primitive at this point.

3. As no great circle on the net passes through its marked point, it is necessary to revolve the overlay until one does.

4. To locate the great circle representing the plane, count off from the primitive on the right side of the net inward along the east-west diameter of the net. Trace in this arc of a great circle.

5. Revolve the overlay back to the original position and check the result by visualization.

3.3.3 Procedures for plotting poles

To plot the pole:

a. Revolve the north mark on the overlay at the strike angle anticlockwise

b. From the east point of the net, count off the dip along east-west diameter

c. Trace in the great circle

d. Count off 90^0 from the plane along the east-west line and put a point labeled "P" which   is the pole

e. The line marking the great circle that represent the plane is cleaned up completely as point "P" (pole) represent the plane.

3.3.4. Procedure for using counting Net

1. Superimpose the point diagram (collection of plotted pole) and a second tracing sheet on the counting net. At the center of each hexagon, the total number of points within that hexagon is written for the main body of the

diagram. The second tracing sheet would be rotated until many of the points are concentrated within one or two hexagons. These will be a number at the center of each overlapping hexagon. For parts of the diagram without any point in the hexagon, it is left blank rather than adding zero.

2. At the periphery of the net, the points in each half hexagon on one side of the net are combined with the complementary half on the opposite side and this number is written on both sides of the net.

3. Points at the ends of the spokes are counted using the complementary half circle.

3.3.5 Procedure for contouring the points

1. To facilitate comparison of diagram with different numbers of total points, contours are drawn in percentages of the total points per 1% area of the net represented by each hexagon. The counting net is a system of triangles, each six of which makes a hexagon. Each hexagon represent one percent (1%) of the total number of hexagons that constitute the whole counting net. Therefore, the number posted during the counting must be converted to percentages.

2. Within the main body of the diagram, contours of equal density are drawn.

3. For contour lines the perimeter, the counts along the edges are used when contour line intersects the primitive is must reappear exactly 180^{0} opposite.

4. When the preliminary contouring is complete, several modifications may be made in order to improve the diagram.

5. The maximum, found during the counting may not be the true maximum of the diagram. The point of greatest concentrations can be found, by returning the point diagram to the counting net using the central 1% circle, adjust the diagram until the largest number of points lies within one or two hexagon.

6. All the contour lines may be eliminated and the spacing may be in 2, 4, 8, 12, and etc.

7. The area of maximum concentration is often completely blackened. Although, unnecessary patterns may be used for the areas of lesser concentration. Particularly effective are stipple pattern graded so that the areas of greater concentration have a denser appearance. Line patterns detract from the visual effect of the diagram and should be avoided.

CHAPTER FOUR

MACROSCOPIC AND MICROSCOPIC STUDIES

The mapped portion is underlained by rocks of the Basement complex. The rock types are granites, gneisses, and migmatites. In some few positions tholeitic Basalt were observed. From the samples I obtained in the field, I selected five of them and named them sample A.B.C.D. and E. these are the samples that are to be discussed here in this chapter.

From thin section studies, the mineralogy generally include quartz, microcline, orthoclase, plagioclase, biotite, chlorite and opaque minerals.

4.1 Macroscopic study of sample A

Plate 1 show that sample A is a fine grained rock, it is grey in colour, and it possesses obvious bandings which aligns parallel to each other. These obvious parallel bands otherwise known as gneissification are the reason to suggest that sample A is a "gneissic rock". The mineralogy of this sample observed in the field includes feldspars, quartz and biotite. It covers about 40% of the total portion.

Plate 1, Photograph of Hand sample "A"

4.1.1 Microscopic study of sample A (plate I)

The rock sample in thin section consist of muscovite/chlorite, Orthoclase feldspars as magacryst and Biotite forms the groundmass.

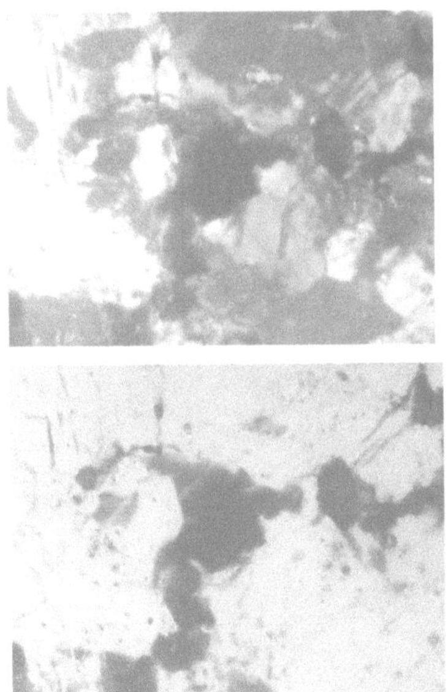

Plate 2a, Photomicrograph of sample A
under cross polarized light

Plate 2b, Photomicrograph of sample A
under plane polarized light

B=Biotite, M=Muscovite, Q=quartz, F=Feldspars, C=Chlorite, O=Orthoclase,

- The colours observed under cross-polarised light include greenish blue, pinkish grey, light grey, dark grey and Brown colours which disappear upon rotation of the stage.

- Due to the fact that muscovite is a transparent mineral, some of the characteristics of minerals lying under it can be seen; therefore Biotite was seen beneath Muscovite forming bands of chlorite. On another position of this slide, the muscovite was clearly seen where it weathers to chlorite.

- The orthoclase crystals under cross polarized light are subordinate to crystals of microcline. And the Quartz crystals under cross polars have anhedral crystals and forms irregular boundaries with other minerals (crystals).

- Pleochroism under plane polarized light, changes from pinkish grey to Dark grey

- Carlsbad twinning was observed.

- The crystals generally exhibit low relief under plane polarized light.

4.1.2 Statistics of sample "A" (plate I)

Table 2, Statistics of sample "A" **(plate I)**

SLIDE POSITIONS	1	2	3	4	5	6	7	8
MUSCOVITE	20	35	25	10	5	20	0	-
CHLORITE	15	0	10	5	-	-	10	20
FELDSPARS	45	55	30	-	10	25	15	10
BIOTITE	40	25	30	10	40	30	10	25

Statistical calculations

Muscovite   =   115

Chlorite   =   60

Feldspars   =   190

Biotite   =   210

        Total   <u>575</u>

Muscovite   =   115/575 x 100   =   20% Muscovite

Chlorite   =   60/575 x 100   =   10.43% Chlorite

Feldspars   =   190/575 x 100   =   33.04% Feldspars

Biotite   =   210/575 x 100   =   36.52% Biotite

4.2.1. Macroscopic study of sample B

Sample B is a medium grained rock, in hand specimen looks like Bauchite only that the grains are not as coarse as that of Bauchite. The fresh sample is grey to dark and greenish especially under the sun, and a bit darker than sample A. Observable minerals in hand specimen are feldspars (plagioclase) biotite and quartz, it covers about 30% of the total area of the portion.

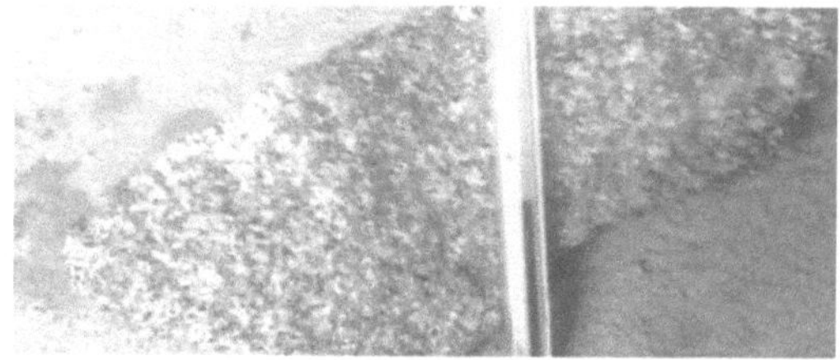

Plate 3, Photograph of Hand Sample "B"

4.2.2 Microscopic study of sample B (plate II)

The rock sample (plate II) under thin section consists of muscovite, microcline, plagioclase, quartz, and Biotite. The Brown colours (Biotite) and the grey colours have equal proportions mixed up as observed.

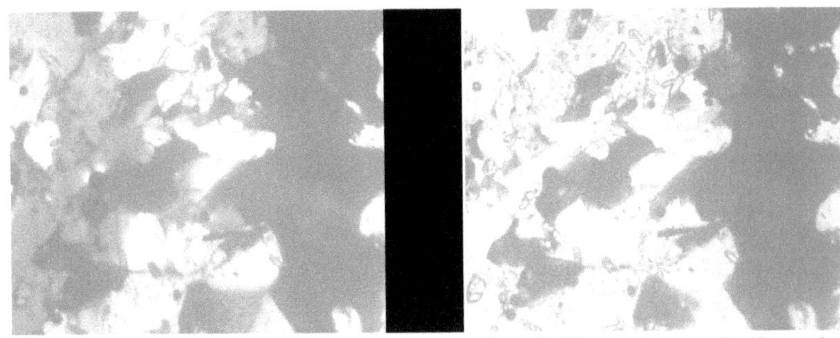

Plate 4a, Photomicrograph of sample B under cross polarized light

Plate 4b, Photomicrograph of sample B under plane polarized light

B=Biotite, M=Muscovite, Q=quartz, F=Feldspars,

- The interference colours observed under cross polarized light include blue, brown, pink, light grey, and Dark grey.

- Under cross polarized light, albite twinning was observed

-The chlorite crystals observed under cross polars are alteration product of biotite

-Under cross polars crystals exhibit wide range of interference colours. Crystals margin are irregular and relationship with other crystals is interlocking.

- Under plane polarized light, no cleavage was observed, pleochroism changes from light grey to dark grey which were the isotropic minerals

- The microcline crystals are not observable under plane polars,

- The crystals observed have low relief under plane polars, straight boundries and crystals   of quartz cracked and inclusion of opaque minerals is common.

4.2.3 Statistics of sample B (plate II)

Table 3, Statistics of sample B (plate II)

SLIDE POSITIONS	1	2	3	4	5	6	7	8
CHLORITE	-	-	-	-	-	-	-	-
QUARTZ	30		5	15	40	25	5	5
FELDSPARS	20	30	10	35	60	40	25	15
MUSCOVITE	20	5	-	-	-	-	15	10
BIOTITE	30	20	35	5	30	35	15	40

Statistical calculations

Quartz   =   125

Feldspars   =   235

Muscovite   =   50

Biotite   =   210

Total   620

125/620 x 100 =   20.16% Quartz

235/620 x 100 =   37.90% Feldspar

50/620 x 100   =   8.06% Muscovite

210/620 x 100 =   33.87% Biotite

4.3. Macroscopic study of sample C

Sample C was obtained from a pegmatite dyke, the sample is therefore pegmatitic, major mineralogy of which were quartz and feldspars (plagioclase and orthoclase). But under thin section some percentage of Biotite were counted which were believed to be originated from the host which was Biotite granite rock, it is biotite granite hosting pegmatite. The sample was light grey to pinkish colour and had pegmatitc texture just like other pegmatite.

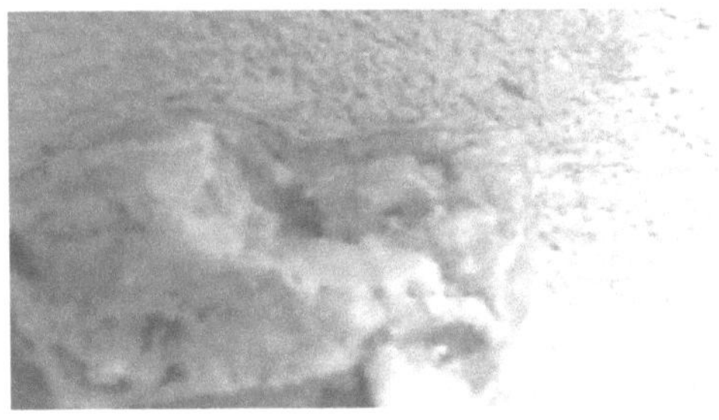

Plate 5, Photograph of Hand Sample "C"

4.3.1 Microscopic study of sample C. (plate III)

Under thin section, plate III shows limited interference colours as light grey, dark grey and brown colours which but disappear upon rotation of the microscope stage. Some pinkish colours were also observed.

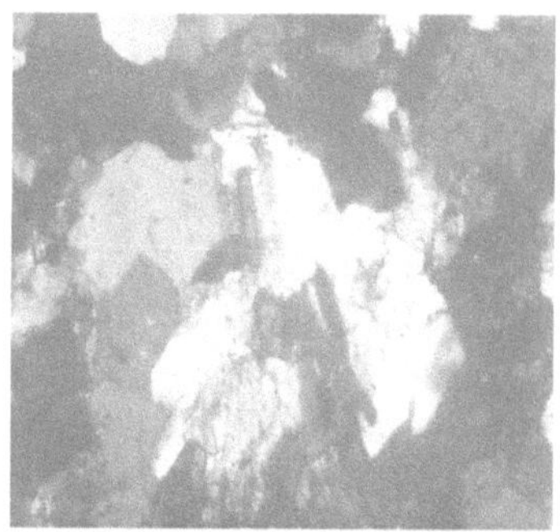

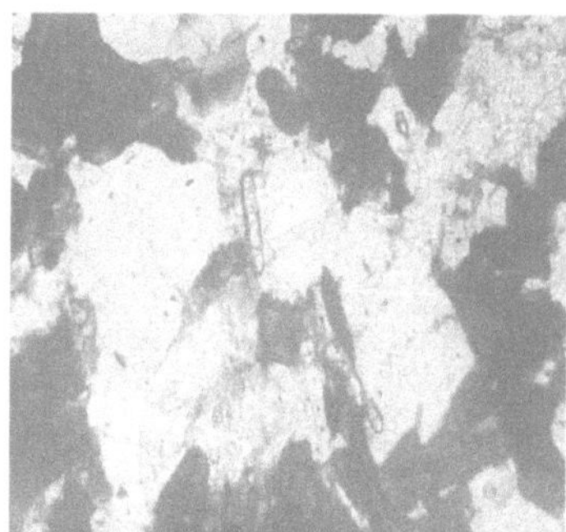

Plate 6a, Photomicrograph of sample C Under cross polarized light

Plate 6b, Photomicrograph of sample C under plane polarized light

B=Biotite, Q=Quartz, P=Plagioclase, O=Orthoclase,

- Albite twinning, was observed under cross polarized light

- Crystals are of low relief and of irregular shape. And had Imperfect cleavage.

4.3.2 Statistics of sample "C" plate III

Table 4, Statistics of Sample C (plate III)

SLIDE POSITIONS	1	2	3	4	5	6	7	8
BIOTITE	30	20	10	25	10	1	15	15
QUATZ	0	30	5	10	10	15	5	-
PLAGIOCLASE	50	55	45	40	35	-	30	40
ORTHOCLASE	0	0	10	-	15	-	20	5

Statistical calculations

Biotite   = 125

Quartz   = 75

Plagioclase =275

Orthoclase =50

 Total =252

125/525 x 100   =23.81% Biotite

75/525 x100   = 14.29% Quartz

275/525 x100   =52.38% plagioclase

50/525 x100   =9.52% Orthoclase

4.4.0 Macroscopic study of sample D

These cover about 45% of the mapped Area. The particular sample is however fine grained; some of the biotite granite gneisses were medium to course grained rocks. Grey to brownish in colour (hand specimen) with well developed foliation defined by planer arrangement of tabular feldspar crystals and biotite shows lineation fabric. They form steeply high outcrops which were jointed and highly fragmented; they are cross cutted by pegmatite/aplite dykes and quartz veins. Fine grained gneisses occurring as cobbles and boulders are observed but relationship with the other gneisses is not well defined. The mineralogy observed include muscovite, biotite, feldspar, and quartz.

Plate 7, Photograph of Hand Sample "D"

4.4.1   Microscopic study of sample D (palte IV)

Major minerals identified under thin section include biotite, muscovite, plagioclase and quartz. Biotite crystals Forms the ground mass and feldspars and quartz were the megacryst.

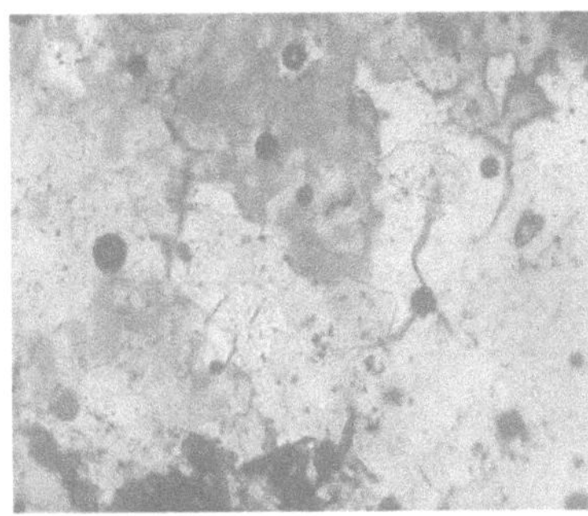

Plate 8a, Photomicrograph of sample D
Under cross polarized light

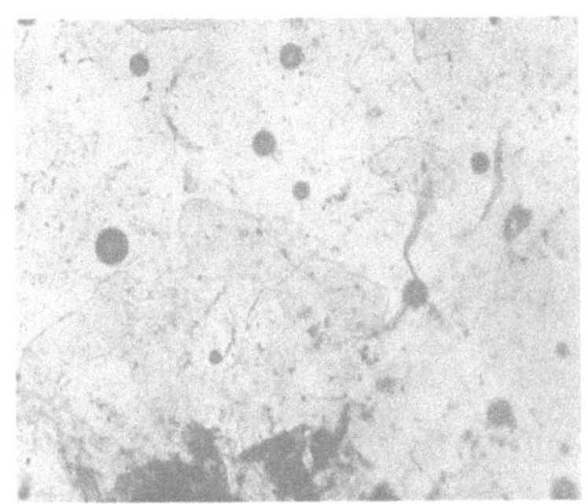

Plate 8b, Photomicrograph of sample D
under plane polarized light

B=Biotite, M=Muscovite, F= Feldspars, Q=Quartz

-   The interference colours observed under cross polarized light include light grey, dark grey, bluish colour and a brown colour which disappear upon microscope stage rotation

- Pleochroism; the bluish colour changes from bluish to light grey and then disappear.

- It is of low relief and crystals have irregular boundries with each other.

- None of the twinning is observed and cleavages were not recognized.

4.4 Statistics of sample "D"

Table 5, statistics of Sample D (plate IV)

SLIDE POSITIONS	1	2	3	4	5	6	7	8
MUSCOVITE	55	95	30	-	40	35	15	30
BIOTITE	25	-	15	-	35	35	40	25
FELDSPARS	15	-	20	-	-	20	30	10
QUARTZ	0	5	10	-	20	10	5	0

Statistical calculations

Quartz   =   50

Feldspars   =   95

Muscovite   =   300

Biotite   =   175

Total   620

50/620 x 100   =   8.06% Quartz

95/620 x 100   =   15.32% Feldspar

300/620 x 100 =   48.39% Muscovite

175/620 x 100 =   29.84% Biotite

4.5.0 Macroscopic study of sample E

This covers less than 20% of the portion, is medium grained granite embedded within gneisses, occurs along with sample D. named above. The medium grained granite occurring between the gneissic granite is much lighter than the hosting granite gneiss which is Darker and fine grained

Plate 9, Photograph of Hand Sample "E"

4.5.1 Microscopic study of sample E. (plate IV)

The major minerals observed were quartz, feldspars, muscovite and biotite crystals.

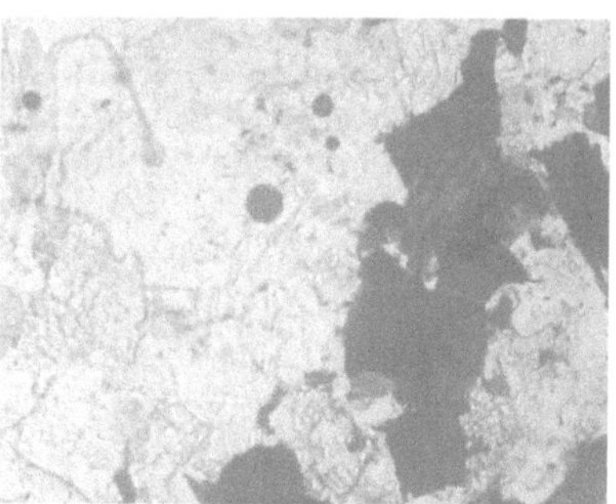

Plate 10a, Photomicrograph of sample E
Under cross polarized light

Plate 10b, Photomicrograph of sample E
under plane polarized light

B=Biotite, Q=Quartz, F= Feldspars, O=Orthoclase,

-Interference colours were blue, greenish, light grey and Dark grey.

-Two types of brown colours were observed, one of the brown colour disappear upon rotation while the other brown colour is a fixed brown colour which always maintain its brownish appearance regardless of the rotation

-Pleochroism; bluish colour to light grey upon microscope stage rotation.

-Albite twining was observed under cross polarized light

-Under plane polarized light, it was observed that crystals form irregular boundries with each other and were of low relief.

-Magnification (eye piece x objective lens) was change from x 10 to x 40 to determine the cleavage, and a perfect one directional cleavage was observed

4.5.2 Statistics of sample "E"

Table 6, Statistics of sample "E"

SLIDE POSITIONS	1	2	3	4	5	6	7	8
MUSCOVITE	-	-	-	-	-	-	-	-
BIOTITE	30	20	30	15	20	-	35	40
FELDSPARS	50	60	65	50	45	20	30	35
QUARTZ	15	5	0	0	-	5	10	5

Statistical calculations

Quartz   =   40

Feldspars   =   355

Biotite   =   190

Total   = 585

40/585 x 100   =   6.84% Quartz

355/585 x 100 =   60.68% Feldspar

190/585 x 100 =   32.48% Biotite

4.6 Structural Geology

The area mapped which forms part of the Nigerian basement complex is Affected by the pan-African (about 550ma) and older orogenesis. However the pan African orogeny tends to erase the imprints of the other orogenies in the area, therefore, its effect overshadowed those of the older orogenies in the area.

Based on field observations, the geologic structures observed in the area include joints, foliations, minor folds, ghost shist structure and dykes & veins.

4.6.1 Joints

These are the most prominent of all the tectonic structures found in the area, however, they were only observed on the gneisses. They occur mostly as tight fractures, however, in many cases fissured joints were observed due to weathering. Vegetation usually grows along the trend of such fissures which suggest that they are water conduits (Figure 11). Most of the joints were in multiple of twos or threes that is joint set. The joints show variable trends from NW-SW. The NE-SW trends were dominant and therefore shows masking effect of the Pan-African orogeny on the older granites (striking NW-SE).

Plate 11, Tree growing in a Joint within Granite Gneiss

4.6.2 Lineation (Foliation)

Foliation occurs in all the rock units, they were more conspicuous on the gneisses where foliation is represented by planar arrangement of feldspar porphyroblast. Sometimes biotite also exhibits the foliation fabric. The foliation may be as a result of stresses produced by tectonic forces that have affected the rocks. Provided the rocks are ductile (due to tectonically generated heat) the stresses produced a permanent distortion in the rocks and together with chemical alteration of the rocks, can rotate and change shape of the grains making up the rocks so that their long dimensions is turned away from the direction of greatest shortening of the rock. The foliation here is secondary (i.e. resulting from tectonic activity). Foliations in the basement rock are widespread and trend mainly between N-S and NE-SW (oyawoye 1970). Lineation occurs as linear arrangement of biotite, elongated quartz and feldspar. They are conspicuous within the gneisses and migmatites more especially near bands (i.e. the bands of alternating mafic and feldsic minerals).

Plate 12, foliations within Granite Gneiss

4.6.3 Minor Folds

This occurs mainly on the gneisses outcrop and it was only observable on the outcrop. Feldspar Porphyroblast forms tight to isoclinals folds with sharp axial planes. Most of the fold can be described as similar folds because the lower and upper surfaces of the folds were identical in shape. These structures were observed on low lying outcrops.

Plate 13, minor folds within Granite Gneiss

4.6.4 Dykes (Pegmatite/Aplite) and vein (Quartz)

Pegmatite dykes ranging from about 10cm-40cm were common within the area they have cross cutting relationship with the rock units, however, they were not continues. Therefore, they were injections or permeations into the main rock units most probably along weak zones and their trend is variable from EW, NW to NE. they were mostly weathered, (Especially the feldsparthic materials); mineralogically composed of quartz and feldspars (i.e. simple pegmatite)

Plate 14, Pegmatite dyke hosted by Biotite Granite

Quartz veins were common and they range in width from about 2cm to 5cm. Both structures were mostly weathered and in most cases were only observed on boulders.

Plate 15, Quartz Vein within Biotite Granite

4.7 Structural analysis

Problems involving angular relationships of lines and planes are best resolved using stereographic projections. It employed using Schmidt net (figure 4). Structures could be analyzed using the S-pole diagram where points of the measured planes are plotted as pattern of points.

The S-pole diagram is chosen because of the size of the data. This implies that each plane is reduced by 2-diamentsion and is plotted as a point located 90 away from a dipping plane. Table below shows the 73 strikes and dip (attitude of foliation planes) readings measured directly from the field using geologic compass.

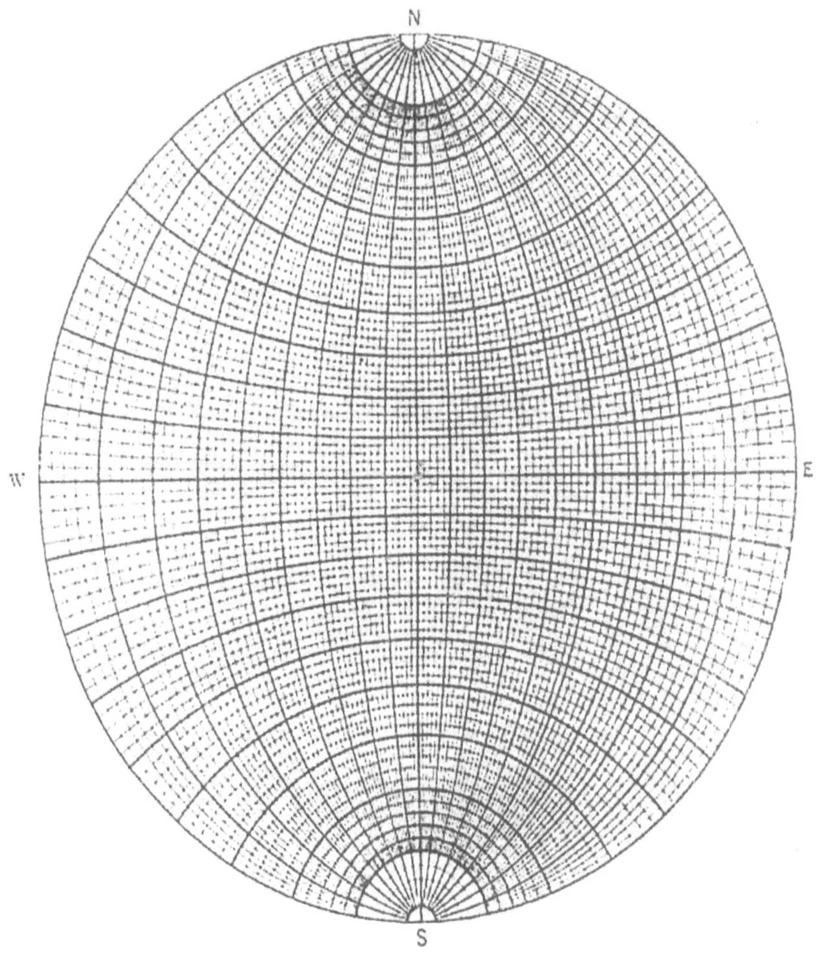

Figure 4, Schmidt net

4.7.1 Point or pole Diagram

A point diagram was drawn using the S-pole method on a tracing paper superimposed on the Schmidt net. After the plotting of the point diagram, the densities of these points are then considered because they are the target.

Attitude readings for migmatites

	STRIKE	DIP			STRIKE	DIP
1	N354^{0}W	42^{0}W		18	S148^{0}E	43^{0}W
2	N352^{0}W	46^{0}W		19	S120^{0}E	28^{0}S
3	N326^{0}W	40^{0}W		20	S100^{0}E	30^{0}S
4	N344^{0}W	46^{0}W		21	N80^{0}E	20^{0}S
5	N282^{0}W	30^{0}W		22	S135^{0}E	22^{0}S
6	N292^{0}W	34^{0}W		23	S143^{0}E	47^{0}S
7	N308^{0}W	40^{0}W		24	S110^{0}E	40^{0}S
8	N285^{0}W	30^{0}W		25	S183^{0}W	50^{0}E
9	N330^{0}W	38^{0}W		26	S170^{0}E	44^{0}W
10	N332^{0}W	40^{0}W		37	S240^{0}W	30^{0}W
11	N344^{0}W	40^{0}W		38	S190^{0}E	40^{0}W
12	N40^{0}E	40^{0}W		29	N68^{0}E	38^{0}S
13	N342^{0}W	27^{0}W		30	S145^{0}E	37^{0}W
14	N348^{0}W	30^{0}W				
15	N320^{0}W	22^{0}W				
16	N290^{0}W	22^{0}W				
17	S220^{0}W	20^{0}W				

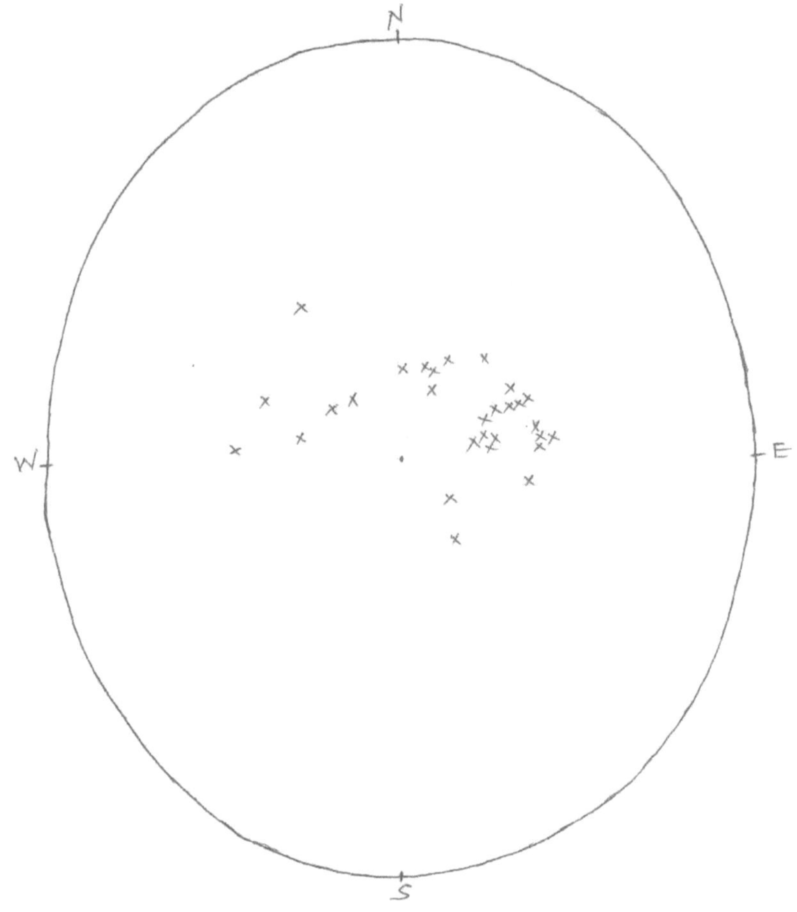

Figure 5, S-Poles diagram for Migmatites

	STRIKE	DIP			STRIKE	DIP
1	N310^{0}W	19^{0}W		16	N322^{0}W	32^{0}E
2	N340^{0}W	30^{0}W		17	N336^{0}W	20^{0}E
3	S200^{0}W	38^{0}W		18	N10^{0}E	20^{0}W
4	S90^{0}E	38^{0}S		19	S96^{0}E	34^{0}W
5	S150^{0}E	20^{0}W		20	N350^{0}W	27^{0}W
6	S160^{0}E	40^{0}W		21	S171^{0}E	16^{0}W
7	S170^{0}E	30^{0}W		22	N360^{0}E	20^{0}W
8	N350^{0}W	38^{0}W		23	S98^{0}E	22^{0}W
9	S220^{0}W	20^{0}W		24	N346^{0}W	20^{0}E
10	N40^{0}E	28^{0}W		25	N26^{0}E	30^{0}W
11	S170^{0}E	30^{0}W		26	N36^{0}E	20^{0}W
12	N319^{0}W	30^{0}E		27	N30^{0}E	25^{0}W
13	S258^{0}W	31^{0}E				
14	N355^{0}W	32^{0}W				
15	N356^{0}W	24^{0}W				

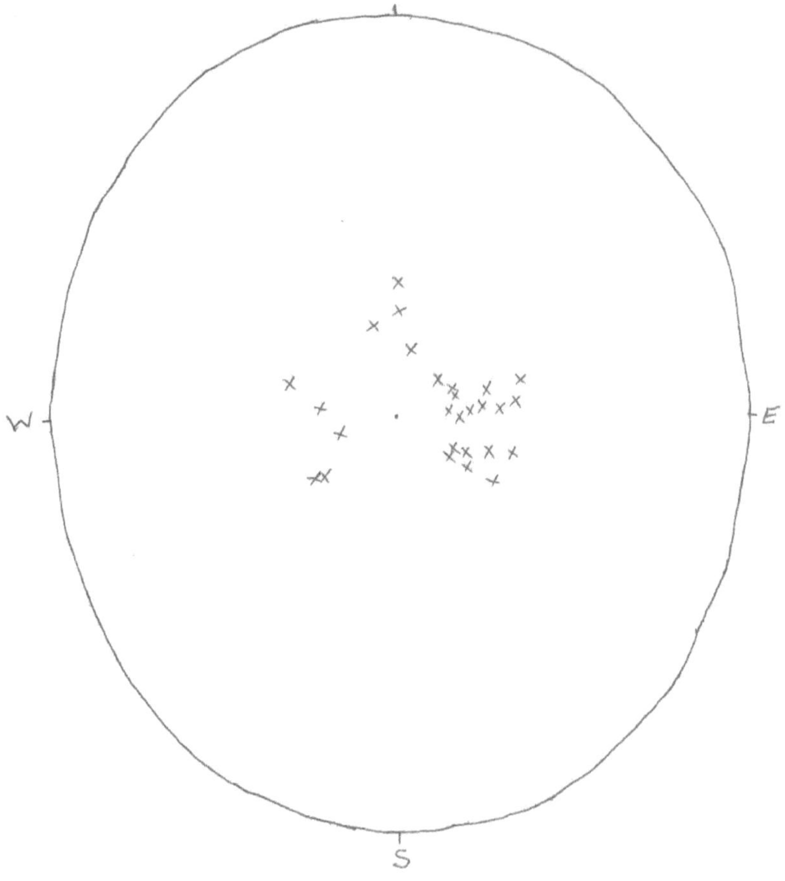

Figure 6, S-Pole diagram for the Gneisses

Attitude readings for Biotite Granite

	STRIKES	DIP			STRIKE	DIP
1	N284^{0}W	51^{0}W		11	S252^{0}W	41^{0}W
2	N270^{0}W	44^{0}W		12	N295^{0}W	12^{0}W
3	N288^{0}W	30^{0}W		13	N286^{0}W	35^{0}W
4	N294^{0}W	41^{0}W		14	N284^{0}W	32^{0}W
5	N286^{0}W	46^{0}W		15	N274^{0}W	37^{0}W
6	N270^{0}W	23^{0}W				
7	S265^{0}W	42^{0}W				
8	S245^{0}W	30^{0}W				
9	S248^{0}W	34^{0}W				
10	N274^{0}W	32^{0}W				

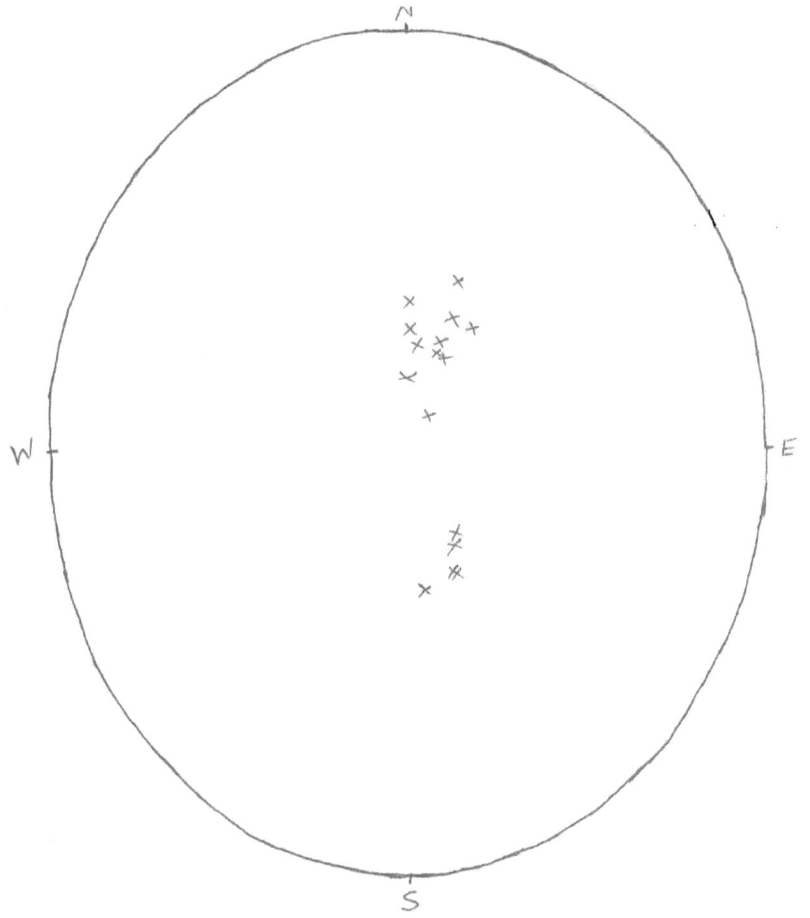

Figure 7, S-Pole diagram for Biotite Granite

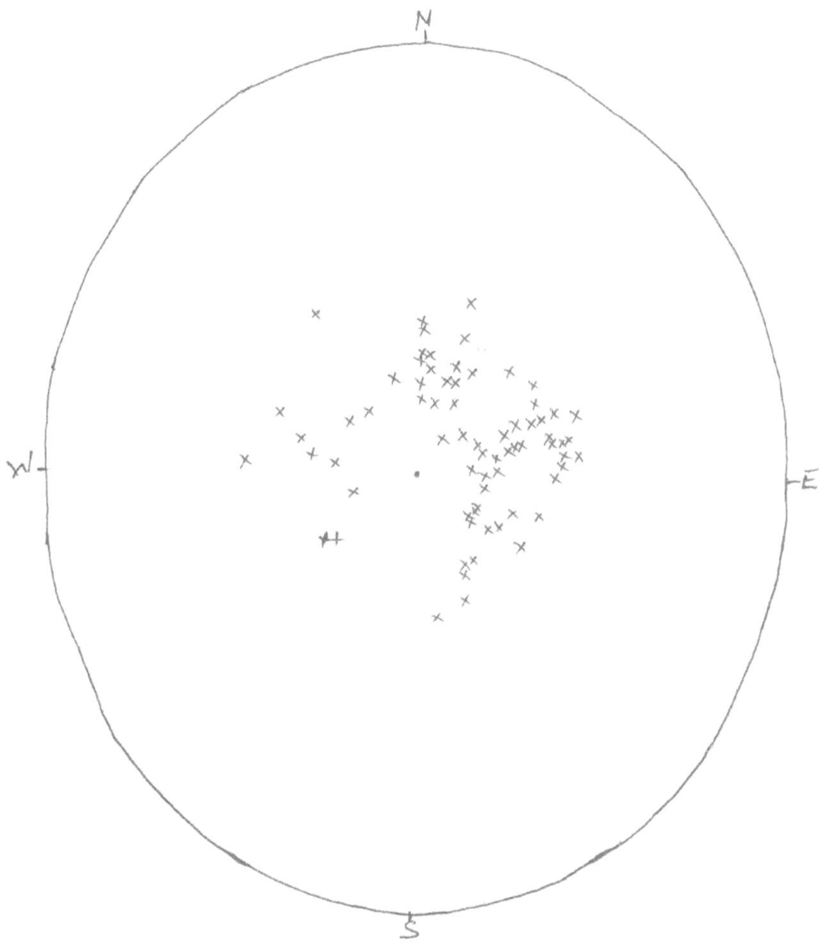

Figure 8, S-Pole diagram for all the 73 attitude readings

Counting net: The method we used here is one of the simplest yet devised and it applies reasonably well to all situations. A special counting net (Fig 9) is required which is completely sub – divided into small triangles, six of the triangles form a hexagon area to equal one percent (1%) of the total area of the net. In addition, to ease the use, this counting net has the advantage of a fixed relationship between the total numbers of points and the centered density. In our point diagram (poles) (Fig 10), the 73points are well spread and some of them fell within 18 hexagons (A to R). Hexagon I, O and R are adjacent and they have the highest congestion of points. They are aligned in the general direction of spread of the point structures of points respectively. (See figure 10)

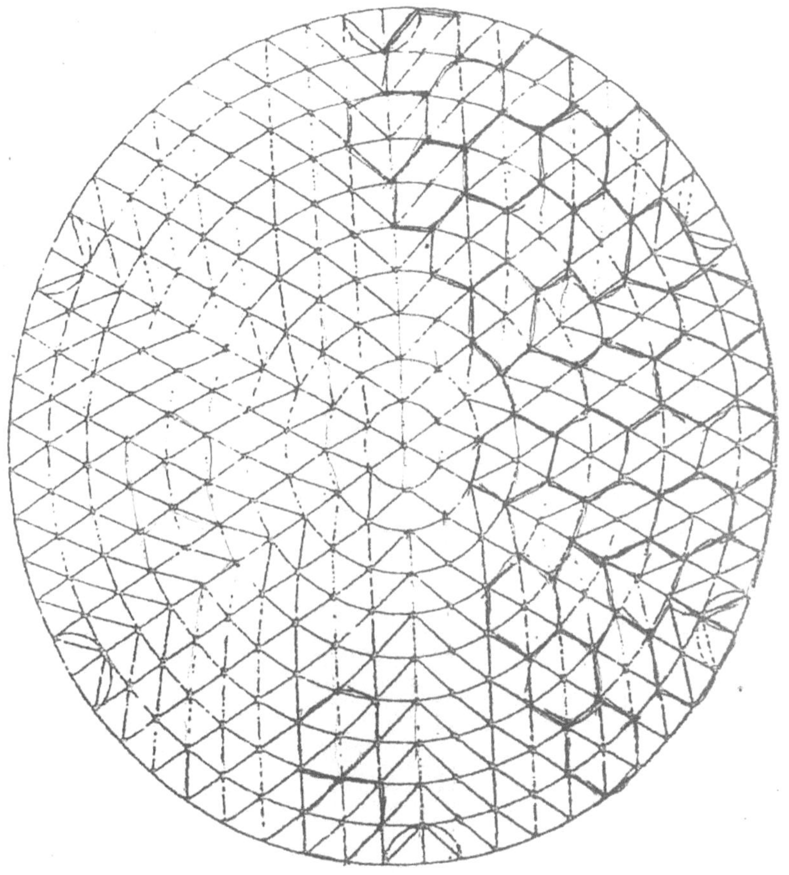

Fig 9 Counting net

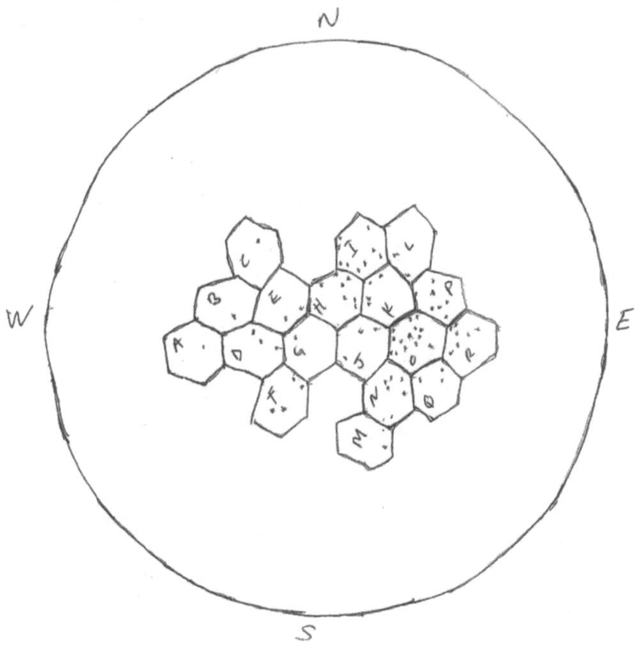

Figure 10, Points plotted within Hexagons

Contouring of point Diagram

Hexagons I, O and R are chosen for the counting because they have the highest point's densities in the general direction of spread of the point. This is achieved by rotating the overlay until one or two hexagons host the highest congestion of points (Poles). From the foregoing we know that 73 plotted points on the counting net is equivalent to 100% total area of the counting net, implying that each plotted points is equivalent to the total points plotted. That is considering the conversion of points on the counting net to 100%. The counting is done by linking points of equal densities within the three chosen hexagons (I, O and R) with the highest congestion of points. The contour interval is in the format; 2, 4, 8, 12 etc. to avoid congestion of the contour lines that could results in difficult interpretation. This is done on a separate tracing sheet (a 3rd overlay sheet superimposed). The central portion of the congested points is always colored black, to show the shape of the girdle which depicts extent of deformation.

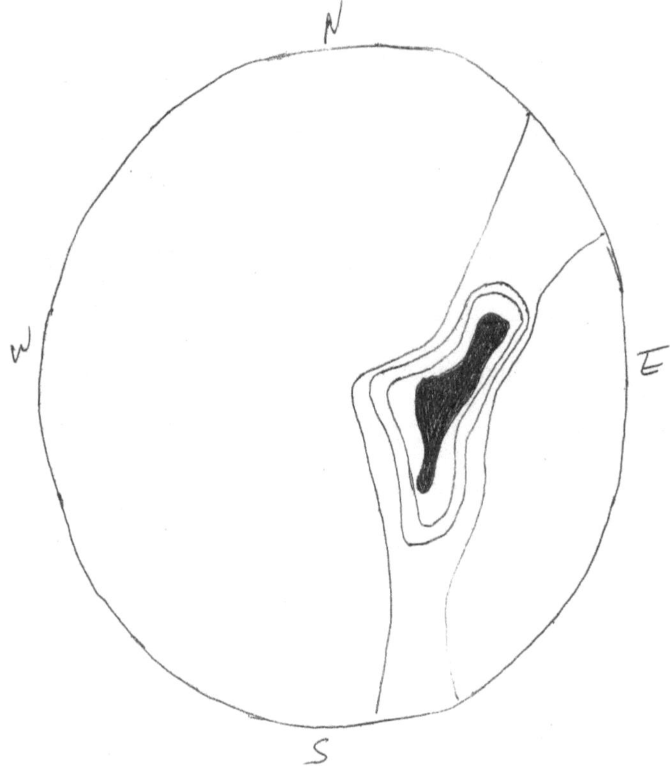

Figure 11, Contoured diagram

CHAPTER FIVE

5.0 RESULTS AND SUMMARY

5.1 Petrography

The selected samples were named sample A, B, C, D and E, the work done on each of these include field description, petrographic analysis and the statistics of the minerals. Based on these therefore, the samples were named as.

-Sample A- Granite Gneiss

-Sample B- Biotite Granite

-Sample C- Pegmatite

-Sample D- Biotite Granite Gneiss

-Sample E- Biotite Granite hosting pegmatite.

5.1.2 Interpretation of diagrams (structural analysis)

The girdle (Fig.11); the girdle is defined as the grouping of points in a band along a great circle. In this work the girdle is much more suitable for interpretation of the pattern of structures perhaps the most important advantage is that S-pole diagrams is based on a statistically generated valid coverage of the structure which gives information concerning the shape of the deformed or folded surface and could be used to tell about the degree of metamorphism (deformation)

Our pattern of contour here shows a shape of medium folding due to medium spread of points and a medium stretching of the girdle. This therefore coincides with medium grade metamorphism with a relatively medium pressure. Generally we may say that the planes of foliation here are gently folded. (See figure 12 below).

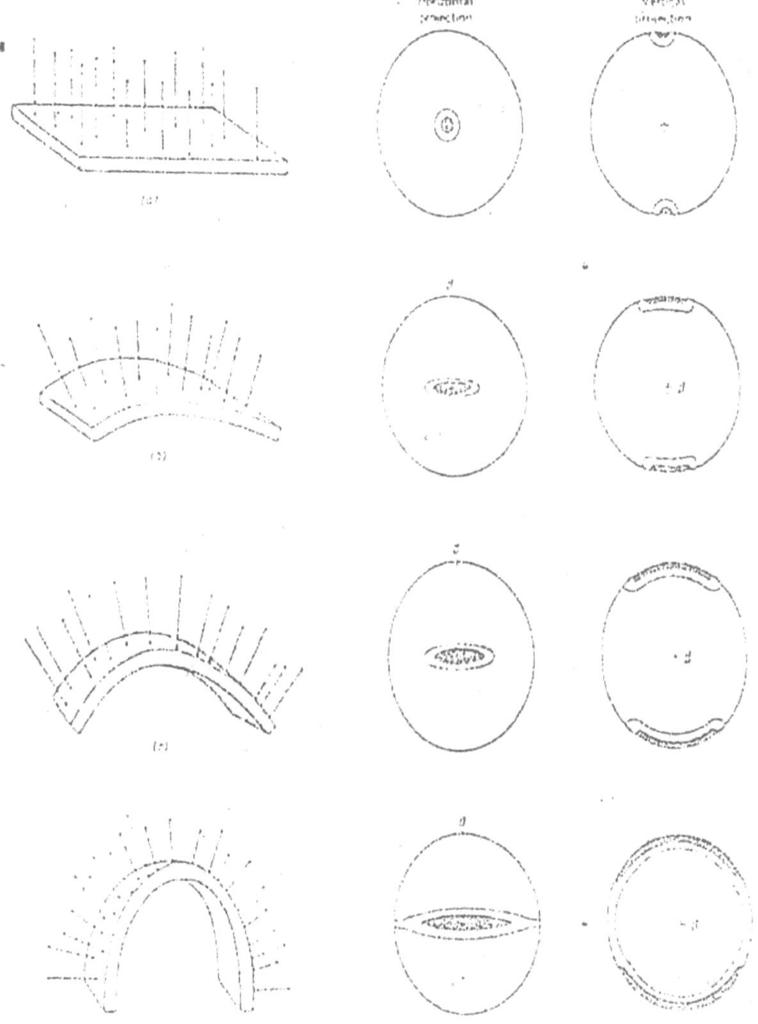

Fig. 1│ Development of S-pole during folding

Conclusion

In conclusion, the structural analysis done above indicates that the rocks in the area were subjected to medium grade metamorphism (see page 55-56), from field work. Rocks in the area were mainly migmatite, biotite granite and gneisses. The gneisses which were distinguished based on texture and field relations belong to the older granite suite of the Pan-African Carter et al., (1963). And Macleod et al., (1970). From my thin-section, the rocks are granitic in composition because they are exclusively quartz, orthoclase, microcline, plagioclase, biotite, chlorite and the opaque minerals. The gneisses and the migmatite are polycyclic having undergone different episodes of tectonism and metamorphism to some varying degrees. Therefore the structural analysis, thin-section and the field data all supplement to each other.

Joints and foliation which were of tectonic origin were abundant in the rocks of the area and their trends were variable (E-W, NW-SE and NE-SW). The NE-SW trends were dominant and thus, among others indicate the masking effect of the Pan-African orogeny in the area. Other structures include ghost schist structure, minor fold, dykes and veins.

The alluvial deposit and the fractured rocks at depth forms the main aquifer systems in the area, and fractured layer at depth forms the more important aquifers. The mapped area contains steeply high insulbergs of the crystalline basement. In some areas (low lying areas) the rocks are not exposed

due to alluvial deposit that covers them. These are believed to be derived from

the weathering of the rocks, and they form fertile soil good for agricultural

purposes.

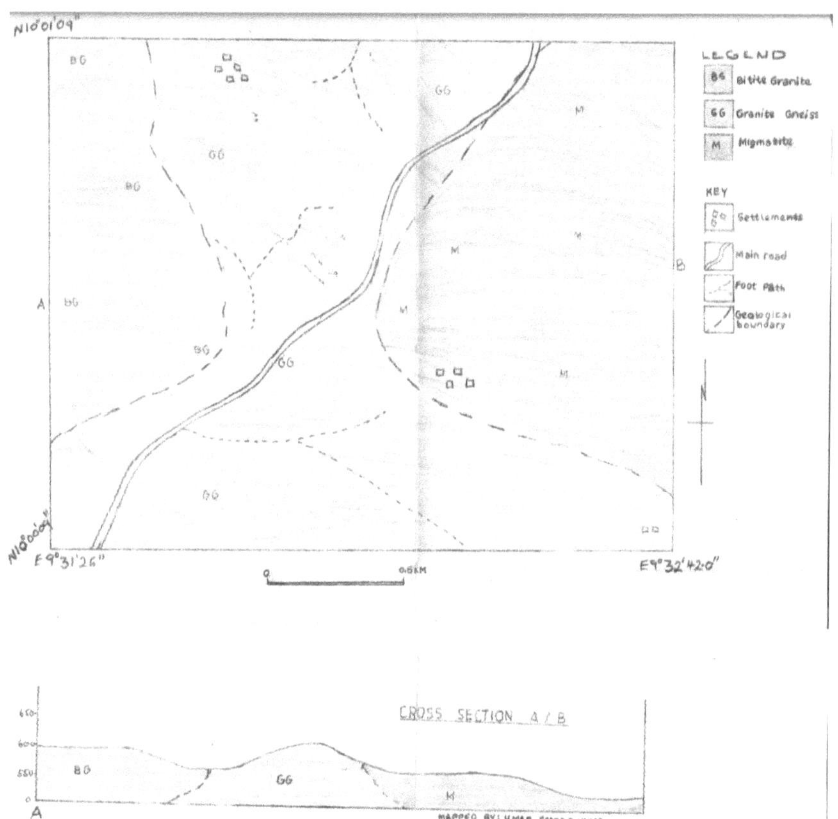

Fig. 13; Geological map of the study area (Kanwara, Dass LGA, Bauchi State-Nigeria)

Recommendations

1. Though no mineral of economic value was encountered, there are still rooms for discovery. Therefore, a more detail work should be carried out in the area with the aim of searching for any existence of such.

2. Firms involve in civil engineering works should be encouraged to explore the aggregate and polished stone potentials of the area, this will go along way in providing jobs for youths of the area thereby alleviating poverty within the local community and that will also provide revenue at least to local and state governments.

3. Most importantly, more boreholes are needed in the area and these boreholes should be strategically located in a way to serve the communities, they may also help in irrigation farming.

References

Ajibade A.C. and Woakes., M. (1976): Proterozoic Crustal Development in the Pan-African Regime of Nigeria., C.A. Kogbe (editor). Geology of Nigeria.

Akpokodje, E.G., (1992). "Properties of some Nigerian Rock aggregate and concrete" *journal of mining and Geology Vol. 28 No. 2.* Nigerian mining and Geosciences Society, Enugu, Nigeria.

Bain A.D.N.,(1926): "The Geology of Bauchi town and surrounding district. Geol. Surv. Of Nigeria, Bull. No. 9

Carter, J.D, Barber, W.M; D.F and Tait, E.A., (1963) "The Geology of part of Adamawa, Bauchi and Borno province in Northeastern Nigeria. Geol. Surv. Of Nigeria, Bull. No 30.

Cooray, P.G.,(1974): "some aspect of the Precambrian of Nigeria:" A review Journal of mining and Geology Vol. 8 No 1 and 2. The Nigeria mining and Geosciences society.

Dietrich, R.V and skinner, B.J., (1979): "Rocks and Minerals" John-Wiley and Sons Inc. Canada USA

Eborall, I.M.,(1989): "intermediate rock from older granite complex of the Bauchi area, northern Nigeria" In C.A. Kogbe (Editor) Geology of Nigeria, Rock view ltd, Jos, pp71-80

Edok-ete-Madilas., (1976): "Hydrogeological and Geophysical investigation for groundwater in Bauchi State Edok-Eter-Mandilas Limited, Bauchi.

Ekwueme, B.N., (1993): "An easy approach to metamorphic Petrology" University of Calabar Press, Calabar Nigeria.

Elueze, A.A.,(1995): "Prospect for sourcing stone-polishing ventures from rocks in the basement complex of Nigeria" Journal of mining and Geology Vol. 31 No.1 Nigerian mining and Geosciences society, Enugu, Nigeria.

Falconer, J.D., (1911): "The geology and geography of Northern Nigeria." Macmillan and co. Ltd, London.

Heinrich, E.W.,(1965): "Microscopic identification of minerals" McGraw-Hill book company; Michigan, USA

Hobbs, B.E.: Means, W.D and Williams, P.F., (1976): "An outline of structural Geology "John Wiley and Sons, New York USA.

Kerr, P.F., (1977): "Optical mineralogy" McGraw Hill Inc. New York USA.

Macleod, W.N.; Turner, D.C; Wright, E.P., (1971): " The Geology of the Jos Plateau" Geol. Suvr. Of Nigeria. Bull. No. 32.

McCurry, P.,(1989): "A general review of the Geology of the Precambrian to lower Paleozoic rocks of Northern Nigeria." In C.A. Kogbe (Editor) Geology of Nigeria. Rock view Ltd Jos pp13-37

Moorehouse, W.W., (1959): "The study of rocks in thin section" Harper and Row, New York. USA

Patricia McCurry., (1989): "A General Review of the Precambrian to Lower PaleozoicRocks of   Northern Nigeria" M.C.Kogbe (editor). Geology of Nigeria, 2nd revised edition.